CONNECTING THE DOTS
HOW IOT IS TRANSFORMING INDUSTRIES

Brad Young

Connecting The Dots: How IOT Is Transforming Industries

Copyright © 2024 by Brad Young

ISBN: 979-8-89298-159-0

TABLE OF CONTENTS

5

PREFACE

The dawn of the Internet of Things (IoT) represents one of the most significant technological shifts of our era. As we transition into a world where virtually every device is interconnected, the implications for industries, businesses, and daily life are profound. This book, **Connecting the Dots: How IoT is Transforming Industries**, seeks to explore the vast potential and the challenges that come with this technological revolution. Whether you are a business leader looking to leverage IoT innovations, a technology enthusiast eager to understand the latest advancements, or a policymaker aiming to navigate the complexities of this new interconnected landscape, this book offers valuable insights and practical knowledge.

INTRODUCTION

The Internet of Things (IoT) is more than just a buzzword; it is a transformative force reshaping the way we live, work, and interact with the world around us. By enabling devices to communicate and share data, IoT offers unprecedented opportunities for innovation across various sectors. From enhancing the efficiency of industrial processes to revolutionizing healthcare delivery and optimizing agricultural practices, IoT holds the promise of a smarter, more interconnected future.

In Connecting the Dots: How IoT is Transforming Industries, we delve into the myriad ways IoT is influencing different industries. We start with the most visible applications in smart homes and then move through industrial IoT, healthcare, agriculture, retail, and more. Each chapter provides an in-depth look at how IoT technologies are being implemented, the benefits they bring, and the potential challenges posed by widespread adoption.

This book also addresses critical issues related to IoT, such as data security, privacy concerns, and the ethical considerations of deploying interconnected technologies. As we navigate the digital transformation, it is essential to remain vigilant about the risks while capitalizing on the opportunities.

Through a combination of expert analysis and real-world examples, this book offers a comprehensive overview of the IoT landscape. By the end, readers will not only have a deeper understanding of how IoT is transforming industries but also a clear vision of the future possibilities and the steps necessary to harness its full potential. Welcome to a world where everything is connected, and the possibilities are endless.

CHAPTER 1: INTRODUCTION TO INTERNET OF THINGS (IOT)

WHAT IS IOT?

The Internet of Things (IoT) refers to the network of interconnected devices that communicate and exchange data with each other through the internet. These devices, ranging from ordinary household items to sophisticated industrial tools, are embedded with sensors, software, and other technologies that enable them to collect and share data. IoT aims to create a smarter environment, where devices can assist humans in making informed decisions, improving efficiency, and enhancing the quality of life.

EXAMPLES OF IOT:

1. **Smart Thermostats:** These devices can learn a family's routine and adjust the temperature automatically, leading to energy savings and increased comfort.

2. **Connected Fleet Management:** In the logistics sector, IoT plays a crucial role in fleet management. By equipping vehicles with GPS and other sensors, companies can track the location, speed, and condition of their assets in real-time. This not only enhances the efficiency of route planning but also reduces fuel consumption and maintenance costs.

3. **Smart Warehouses:** IoT enables the automation and optimization of warehouse operations through the use of sensors and RFID tags. These technologies help in monitoring inventory levels, tracking goods, and managing storage conditions, leading to improved accuracy and faster order fulfillment.

4. **Predictive Maintenance:** IoT devices can monitor the performance and health of logistics equipment, such as conveyor belts and forklifts. By analyzing data from these devices, companies can predict when maintenance is needed, reducing downtime and preventing costly repairs.

5. **Cold Chain Monitoring:** For the transportation of perishable goods, IoT sensors can monitor temperature, humidity, and other environmental conditions in real-time. This ensures that products remain within safe parameters throughout the supply chain, reducing spoilage and ensuring compliance with regulations.

6. **Wearable Health Monitors:** Devices like fitness trackers and smartwatches monitor health metrics such as heart rate, activity levels, and sleep patterns, providing valuable data for managing personal health.

7. **Connected Cars:** Modern vehicles are equipped with internet connectivity that allows for navigation assistance, telematics, and even predictive maintenance alerts.

8. **Smart Home Security Systems:** These systems include connected cameras, doorbell cameras, and motion sensors that provide real-time alerts and remote monitoring capabilities.

9. **Industrial IoT Sensors:** Utilized in manufacturing, these sensors monitor machinery and production

processes, optimizing performance and predicting potential equipment failures before they occur.

BENEFITS OF IOT FOR EVERYONE:

1. **Enhanced Efficiency:** IoT devices automate and streamline processes, reducing the need for manual intervention and increasing overall efficiency.

2. **Improved Health and Wellness:** Wearable health devices and connected medical equipment provide real-time monitoring and early detection of potential health issues, promoting better health management.

3. **Energy Savings:** Smart home devices such as thermostats and lighting systems help reduce energy consumption by optimizing usage patterns, leading to lower utility bills.

4. **Increased Safety:** IoT security systems offer advanced protection for homes and businesses, providing real-time alerts and remote access to surveillance footage.

5. **Convenience:** IoT technology simplifies daily tasks, from remote control of household appliances to

automated reminders and alerts, making life easier and more convenient.

Overall, IoT continues to transform various aspects of our lives, bringing unprecedented connectivity, convenience, and insight into everyday activities.

HOW IOT WORKS

To understand how IoT works, it is essential to grasp the foundational technologies that underpin its functionality. At its core, IoT relies on a network of devices equipped with sensors, actuators, and communication hardware. These devices collect data from their environment, such as temperature, light, motion, or moisture levels, and transmit this information wirelessly to a central system for processing and analysis. This data can then be used to trigger specific actions or send alerts to users, making systems smarter and more responsive.

The technology behind IoT comprises several key components: sensors, connectivity, data processing, and user interfaces. Sensors detect and measure changes in the physical environment, capturing vital data. Connectivity technologies, such as Wi-Fi,

Bluetooth, cellular networks, or low-power wide-area networks (LPWAN), enable devices to communicate with each other and with central servers. Data processing units, often located in the cloud, analyze the accumulated data and derive actionable insights. Finally, user interfaces, such as mobile apps, dashboards, or voice-activated systems, allow users to interact with IoT devices, monitor their status, and control their functions seamlessly. This intricate interplay of technologies empowers the IoT ecosystem, creating a connected world where data-driven decisions enhance daily life.

EVOLUTION OF IOT

The evolution of the Internet of Things (IoT) has been a remarkable journey, characterized by rapid technological advancements and significant shifts in how devices interact with the world around them. Initially, IoT focused on simple, single-purpose devices connected via basic connectivity protocols. However, as technology has progressed, IoT systems have become more sophisticated and integrated, employing advanced sensors, machine learning algorithms, and high-speed connectivity options like 5G. This evolution has had profound impacts on various industries and everyday life.

EXAMPLES OF THE EVOLUTION OF IOT AND ITS IMPACT:

1. **Smart Cities:** Early iterations of connected devices were limited to traffic lights or parking meters. Today, entire cities are connected through IoT, utilizing smart grids, intelligent public transportation, and responsive traffic management systems to enhance urban living.

2. **Healthcare Innovations:** Initially, IoT in healthcare included basic monitoring devices. The evolution now includes advanced telemedicine, remote patient monitoring systems, and even AI-driven diagnostics, vastly improving patient outcomes and healthcare accessibility.

3. **Agricultural Efficiency:** Early agricultural IoT applications were limited to simple weather sensors. Modern smart farming uses IoT devices for precision agriculture, employing drones, soil health sensors, and

automated irrigation systems to optimize crop yields and reduce resource usage.

4. **Retail Transformation:** The retail industry began using IoT with early inventory management systems. Now, smart shelves and RFID technology provide real-time inventory data, while personalized shopping experiences are enhanced through IoT-enabled devices that track customer preferences and behaviors.

5. **Autonomous Vehicles:** Initial connected cars offered basic navigation and telematics. The current evolution has led to the development of autonomous vehicles with IoT integration, enabling real-time data exchange between cars, infrastructure, and the cloud, thereby enhancing safety and efficiency on the roads.

The relentless evolution of IoT continues to push the boundaries of what is possible, ushering in an era of unprecedented connectivity and automation that touches nearly every facet of human activity.

IMPORTANCE OF IOT IN TODAY'S WORLD

The importance of IoT in today's world cannot be overstated, as it revolutionizes the way we live, work, and interact with our environment. By enabling seamless connectivity between devices, IoT enhances efficiency, drives innovation, and fosters new opportunities across various sectors. From healthcare to agriculture, the integration of IoT technology has the potential to solve complex challenges, improve quality of life, and create smarter, more responsive systems.

EXAMPLES OF THE IMPORTANCE OF IOT:

1. **Healthcare Monitoring:** IoT devices in healthcare, such as remote patient monitoring systems and wearable health devices, enable real-time tracking of vital signs. This not only improves patient outcomes through early detection but also reduces the strain on healthcare facilities by allowing patients to be treated at home.

2. **Smart Agriculture:** IoT applications in agriculture, such as soil health sensors and automated irrigation systems, help farmers optimize resource usage, increase crop yields, and reduce environmental impact. This leads

to more efficient and sustainable farming practices that can help feed a growing global population.

3. **Energy Management:** Smart grids and IoT-enabled energy management systems can monitor and optimize energy consumption in real-time. By reducing energy waste and improving efficiency, these systems contribute to lower utility bills for consumers and a reduced carbon footprint.

4. **Enhanced Transportation:** IoT technologies in transportation, such as connected traffic management systems and autonomous vehicles, enhance safety, reduce congestion, and improve travel efficiency. Smart transportation systems can lead to fewer accidents, lower emissions, and a more reliable commuting experience.

5. **Smart Home Automation:** IoT-enabled smart home devices, ranging from thermostats to security cameras, provide homeowners with increased convenience, security, and energy efficiency. These devices can be controlled remotely, allowing users to manage their

home environment from anywhere in the world and ensuring optimal comfort and safety.

CHAPTER SUMMARY

The chapter provides an in-depth exploration of the Internet of Things (IoT) and its transformative impact on various aspects of human life. It begins by explaining the foundational components of IoT—sensors, connectivity, data processing, and user interfaces—that work together to create a seamless, interconnected ecosystem. The evolution of IoT is discussed, highlighting the progression from basic single-purpose devices to sophisticated, integrated systems powered by advanced technologies such as 5G and machine learning. Examples are given to illustrate the profound effects of IoT across different sectors, including smart cities, healthcare innovations, agricultural efficiency, retail transformation, and autonomous vehicles. The chapter concludes by emphasizing the crucial role of IoT in addressing complex challenges, improving quality of life, and driving innovation across various fields in today's world.

CHAPTER 2: SMART HOME DEVICES

SMART THERMOSTATS

Smart thermostats have revolutionized home climate control by introducing advanced technology to manage heating and cooling systems more efficiently. These devices not only allow homeowners to set and control their home temperature remotely via smartphone apps but also learn from user behavior to optimize settings automatically. By integrating with other smart home devices, smart thermostats offer a level of convenience and energy efficiency previously unattainable with traditional thermostats.

EXAMPLES OF SMART THERMOSTATS:

1. **Nest Learning Thermostat:** This popular device learns your schedule and preferences, adjusting the temperature automatically to save energy while keeping your home comfortable. It also integrates with various smart home platforms such as Google Home and Alexa.

2. **Ecobee SmartThermostat:** Equipped with voice control, this thermostat offers room sensors to manage

hot and cold spots effectively, ensuring uniform comfort throughout the house.

3. **Honeywell Home T9 Smart Thermostat:** This device uses smart room sensors to control temperature in multiple rooms, ensuring optimal comfort where it matters most. It also learns your schedule and can be controlled remotely.

4. **Emerson Sensi Touch Wi-Fi Thermostat:** Known for its easy installation, this thermostat provides universal compatibility and features such as geo-fencing to adjust the temperature when you leave or return home.

5. **Lux Kono Smart Thermostat:** This stylish thermostat offers smart features like voice control and seamless integration with home automation systems, making it both functional and visually appealing.

Benefits of Smart Thermostats:

1. **Energy Savings:** By learning your schedule and adjusting temperature accordingly, smart thermostats can significantly reduce energy consumption, leading to lower utility bills.

2. **Remote Control:** With smartphone integration, you can control your home's climate from anywhere in the world, ensuring comfort and convenience at your fingertips.

3. **Enhanced Comfort:** Smart features such as room sensors and automatic adjustments ensure that every room in your home maintains the desired temperature levels.

4. **Integration with Smart Home Devices:** Smart thermostats often integrate seamlessly with other smart home systems, such as voice assistants and security systems, creating a unified and intelligent home environment.

5. **User-Friendly Interfaces:** The intuitive design of smart thermostat interfaces makes managing home climate systems easy and accessible for users of all tech-savvy levels.

HOME SECURITY SYSTEMS

Home security systems have evolved dramatically with the advent of smart technology, providing homeowners with unprecedented levels of control and peace of mind. Modern smart home security systems offer a variety of features that allow remote monitoring, automated alerts, and direct integration with other smart devices in the home. These systems utilize advanced technologies such as artificial intelligence, machine learning, and the Internet of Things (IoT) to enhance safety and security measures, making homes smarter and more secure than ever before.

EXAMPLES OF NEW TECHNOLOGY IN HOME SECURITY SYSTEMS:

1. **AI-powered Security Cameras:** These cameras use artificial intelligence to distinguish between different types of movement (e.g., humans and pets), reducing false alarms and increasing security accuracy.

2. **Smart Doorbell Cameras:** Featuring video recording and two-way communication, these devices allow homeowners to see, hear, and speak to visitors at their door from anywhere.

3. **Smart Locks:** These locks can be controlled remotely, offer keyless entry, and can integrate with other smart home devices for enhanced security and convenience.

4. **Motion Sensors with Facial Recognition:** These advanced sensors can identify known individuals and send alerts for unknown faces, providing an extra layer of security.

5. **Automated Lighting Systems:** Integrated with security systems, these lights can be programmed to turn on at specific times or when motion is detected, deterring potential intruders.

Benefits of Modern Home Security Systems:

1. **Enhanced Safety:** Advanced features like real-time alerts and remote monitoring enable quicker responses to potential threats, enhancing overall home safety.

2. **Deterrence of Crime:** Visible smart security devices, such as cameras and smart lights, act as deterrents to potential intruders.

3. **Convenience:** Remote control and automation of security devices make managing home security effortless and efficient.

4. **Energy Efficiency:** Some smart security devices also help manage energy use, such as automated lighting systems that ensure lights are only on when necessary.

5. **Peace of Mind:** Knowing that your home is protected by the latest security technology provides invaluable peace of mind, whether you're at home or away.

SMART LIGHTING

Advancements in smart lighting technology have transformed how people illuminate their homes, making lights more efficient, customizable, and convenient than ever before. Smart lighting systems can be controlled remotely through smartphone apps or voice commands, allowing homeowners to adjust brightness, color, and schedules to suit their preferences and daily routines. By incorporating features such as automation, energy monitoring, and integration with other smart home devices, smart lighting systems provide an enhanced living experience that conventional lighting cannot match.

EXAMPLES OF SMART LIGHTING:

1. **Philips Hue:** Known for its versatility, Philips Hue offers a wide range of smart bulbs, light strips, and fixtures that can change colors, be scheduled, and controlled via voice assistants like Alexa, Google Assistant, and Siri.

2. **LIFX Bulbs:** These Wi-Fi-enabled bulbs offer billions of color options and do not require a hub, making setup straightforward. They can be controlled through smartphone apps and various smart home platforms.

3. **Wyze Bulb:** A budget-friendly option, Wyze bulbs offer features like dimming and scheduling through their app, along with integration with Google Assistant and Alexa.

4. **Nanoleaf Light Panels:** These customizable light panels can be arranged in different patterns and offer dynamic lighting scenes, making them ideal for creating unique ambiances.

5. **Sengled Smart LED:** These bulbs come with features like motion sensing and can be integrated with systems like Amazon Alexa or Samsung SmartThings for seamless control.

Benefits of Smart Lighting:

1. **Energy Efficiency:** Smart bulbs often use LED technology, which consumes less energy compared to traditional incandescent bulbs, leading to lower electric bills.

2. **Convenience:** The ability to control lighting remotely via smartphone apps or through voice commands adds a layer of convenience to daily life.

3. **Customization:** Smart lighting allows users to adjust the brightness and color of their lights to fit their mood or activity, creating personalized lighting environments.

4. **Security:** Automated lighting schedules can be set to mimic occupancy, deterring potential intruders when the home is unoccupied.

5. **Integration:** Many smart lighting systems seamlessly integrate with other smart home devices, creating a cohesive and efficient smart home environment.

SMART APPLIANCES

Smart appliances have revolutionized the way we manage our daily household tasks, streamlining operations and adding a layer of convenience and efficiency to home management. These appliances encompass a range of household devices that can be controlled remotely, often through smartphone apps, and are equipped with features that adapt to individual user preferences and routines. By leveraging the power of connectivity and automation, smart appliances offer advanced functionality that transforms traditional appliances into interconnected home ecosystems.

EXAMPLES OF SMART APPLIANCES:

1. **Smart Refrigerators:** Equipped with touchscreens, interior cameras, and Wi-Fi connectivity, these refrigerators can manage grocery lists, recommend

recipes, and even notify you when items are running low.

2. **Smart Washers and Dryers:** These appliances offer remote control and monitoring, enabling users to start or stop cycles from their smartphones and receive notifications when laundry is done.

3. **Smart Ovens:** Featuring precise temperature controls and recipe-guided cooking, smart ovens can be controlled via apps and often come with integration for voice assistants like Amazon Alexa and Google Assistant.

4. **Smart Dishwashers:** With features like remote start, cycle monitoring, and energy-efficient settings, smart dishwashers adapt to your cleaning needs and can alert you when maintenance is required.

5. **Smart Air Conditioners:** These units can adjust temperatures based on user preferences and can be controlled remotely to ensure optimal comfort while also conserving energy.

Benefits of Smart Appliances:

1. **Convenience:** Remote control and automation features allow users to manage tasks from anywhere, reducing the need for physical presence and intervention.

2. **Energy Efficiency:** Many smart appliances come with energy-saving modes and the ability to monitor usage, helping to reduce household energy consumption and lower utility bills.

3. **Personalization:** Smart appliances can learn user habits and schedules, tailoring their functionality to optimize performance and cater to individual preferences.

4. **Maintenance Alerts:** Automated notifications for maintenance and troubleshooting help extend the lifespan of appliances and prevent costly repairs.

5. **Interconnectivity:** Seamless integration with other smart devices creates a cohesive smart home environment that enhances overall functionality and user experience.

COSTS AND SAVINGS OF SMART HOME ITEMS

Investing in smart home technology can initially seem like a significant expense; however, the long-term savings and benefits often outweigh the upfront costs. Here's a breakdown of the costs associated with various smart home items and how they can help save money in the long run:

Smart Lighting Costs and Savings

1. **Philips Hue:** These bulbs range from $30 to $50 per bulb, with starter kits costing around $80 to $200. Despite the higher initial cost, Philips Hue bulbs use LED technology, which consumes significantly less power and lasts much longer than traditional incandescent bulbs.

2. **LIFX Bulbs:** Priced between $20 and $60 per bulb, LIFX provides an energy-efficient option with no hub required, simplifying setup and reducing additional costs.

3. **Wyze Bulb:** Starting as low as $8 per bulb, Wyze offers an affordable entry into smart lighting. Dimming and scheduling features enable further energy savings.

4. **Nanoleaf Light Panels:** These panels can cost between $200 and $300 per set, but their dynamic and customizable lighting can replace multiple conventional lighting sources.

5. **Sengled Smart LED:** Costing around $10 to $20 per bulb, Sengled offers added features like motion sensing, helping to reduce energy waste when rooms are unoccupied.

The energy efficiency of smart bulbs, coupled with automation features, can significantly lower utility bills over time. Investing in smart lighting might cost more initially but can lead to considerable savings.

SMART APPLIANCES COSTS AND SAVINGS

1. **Smart Refrigerators:** These can range from $1,500 to $4,000. While expensive, their efficiency and ability to keep track of food can help reduce waste and save on grocery bills.

2. **Smart Washers and Dryers:** Prices range from $700 to $1,500 each. With features like energy-efficient cycles

and remote operation, they can reduce water and electricity consumption.

3. **Smart Ovens:** Typically priced between $1,000 and $3,000, smart ovens feature precise temperature control and energy-efficient cooking modes that minimize electricity usage.

4. **Smart Dishwashers:** Cost ranges from $600 to $1,200. Their efficient settings can lower water and energy usage, translating to savings on utility bills.

5. **Smart Air Conditioners:** With prices ranging from $300 to $600, their ability to optimize temperature settings and adapt to user schedules can result in significant energy savings.

Although the upfront costs of smart appliances can be high, their energy-efficient operations, ability to reduce waste, and tailored performance based on user habits generally lead to reduced utility expenses and prolonged appliance lifespan, ultimately providing substantial cost savings over time.

CHAPTER SUMMARY:

This chapter explored the world of smart appliances, detailing various examples such as smart refrigerators, washers and dryers, ovens, dishwashers, and air conditioners. It highlighted the numerous benefits these devices offer, including convenience, energy efficiency, personalization, maintenance alerts, and interconnectivity. Additionally, the chapter provided a breakdown of costs and potential savings associated with smart lighting and smart appliances, emphasizing that the initial investment can lead to significant long-term savings through reduced utility bills and increased efficiency.

CHAPTER 3: INDUSTRIAL IOT (IIOT) APPLICATIONS IN MANUFACTURING

The Industrial Internet of Things (IIoT) represents a significant advancement in manufacturing, offering unprecedented opportunities for enhancing efficiency, productivity, and decision-making. By integrating smart sensors, advanced analytics, and real-time data collection into manufacturing processes, IIoT allows for seamless connectivity across various industrial systems. This integration transforms traditional manufacturing into a digitally connected ecosystem where machines, processes, and human operators work in harmony to optimize performance and reduce downtime.

APPLICATIONS AND USAGES IN MANUFACTURING:

1. **Predictive Maintenance:** By using sensors to monitor equipment health, IIoT can predict when maintenance is needed, helping to prevent unexpected breakdowns and extend machinery lifespan.

2. **Quality Control:** Real-time data analytics ensure product quality by continuously monitoring

manufacturing processes, allowing for immediate adjustments and reducing defects.

3. **Supply Chain Optimization:** IIoT enables better visibility and control over the supply chain, from raw materials to finished products, ensuring timely deliveries and efficient inventory management.

4. **Energy Management:** Smart sensors track energy consumption throughout the manufacturing process, identifying areas for improvement and helping to implement energy-saving measures.

5. **Automation and Robotics:** IIoT facilitates the integration of automated systems and robots, enhancing precision, reducing human error, and increasing production speed.

6. **Asset Tracking:** Real-time tracking of tools, equipment, and finished goods ensures efficient resource allocation and minimizes losses due to misplaced assets.

7. **Worker Safety:** Wearable devices and sensors can monitor worker conditions and environments, alerting to potential hazards and ensuring a safer workplace.

EXAMPLES AND TIME SAVINGS AND OVERALL COST SAVINGS OF EACH

1. **Predictive Maintenance:**

- **Example:** A manufacturing plant uses IIoT sensors attached to critical machinery to monitor vibrations, temperature, and oil levels. These sensors detect deviations from normal parameters and alert maintenance teams before equipment failure.

- **Time Savings:** Predictive maintenance can reduce unplanned downtime by up to 50%. By scheduling maintenance only when needed, companies can avoid lengthy downtime typically associated with reactive repairs.

- **Overall Cost Savings:** Lowering downtime can save millions annually, depending on the scale of operations. For example, a large automotive manufacturer saved $500,000 per year by implementing predictive maintenance.

2. **Quality Control:**

- **Example:** An electronics manufacturer employs real-time data analytics to continuously monitor the quality of the assembly line. Consistent data evaluation ensures that defective products are identified and rectified immediately.

- **Time Savings:** This proactive control can reduce the time spent on end-of-line inspections by up to 30%, and decrease the need for extensive rework.

- **Overall Cost Savings:** Fewer defective products mean reduced waste and lower costs associated with recalls and reworks. An IIoT-driven quality control system saved one company $750,000 annually by cutting defective rate in half.

3. **Supply Chain Optimization:**

- **Example:** A beverage company uses IIoT to track shipments and inventory levels in real-time. Integrated sensors provide data from supplier delivery to final retail distribution.

- **Time Savings:** Enhanced visibility can cut lead times by 35%, ensuring materials and products are delivered exactly when needed, and reducing idle periods.
- **Overall Cost Savings:** Optimizing the supply chain can lower storage and logistics costs by up to 20%. For instance, one company reduced inventory holding costs by $1 million each year.

4. **Energy Management:**

- **Example:** A textile manufacturer utilizes IIoT sensors across its facility to monitor energy consumption at various stages of production. The data collected helps identify inefficiencies.
- **Time Savings:** Automated monitoring and adjustments can improve energy use efficiency without manual interventions, saving substantial operational time.
- **Overall Cost Savings:** Implementing energy-saving measures can reduce utility bills by up to 15%. A case study showed that a manufacturer saved $200,000 annually after implementing an IIoT energy management system.

5. **Automation and Robotics:**

- **Example:** An automotive parts factory integrates robots into its assembly line, controlled by IIoT solutions to perform repetitive tasks with extreme precision and minimal supervision.

- **Time Savings:** Automation can speed up production processes, boosting throughput by 25-30%.

- **Overall Cost Savings:** Automation lowers labor costs and reduces errors, saving up to $500,000 annually in some industries by improving yield and reducing rejects.

6. **Asset Tracking:**

- **Example:** A logistics firm uses RFID tags and sensors on containers and pallets to monitor their locations in real-time.

- **Time Savings:** Real-time tracking decreases the time spent searching for misplaced assets by up to 40%.

- **Overall Cost Savings:** Minimizing asset loss and improving resource use saves hundreds of thousands each year. For example, a logistics company saved $300,000 annually by preventing asset misplacement.

7. **Worker Safety:**

- **Example:** A construction company deploys wearable devices that monitor workers' vitals and environmental conditions, sending alerts in case of detected hazards.

- **Time Savings:** Immediate hazard detection can reduce incident response times by up to 50%.

- **Overall Cost Savings:** Improved safety can lower insurance premiums and reduce lost workdays due to injuries, leading to savings in the range of $250,000 annually for large operations.

By implementing IIoT in various manufacturing applications, companies can significantly enhance efficiency and productivity while achieving substantial cost savings in diverse operational aspects.

CHAPTER SUMMARY

The integration of Industrial Internet of Things (IIoT) in manufacturing encompasses several key applications that collectively enhance efficiency, productivity, and safety. From predictive maintenance and quality control to supply chain optimization and energy management, IIoT technologies facilitate real-time data analytics and automated responses. By embedding sensors and smart devices throughout the production process, manufacturers can prevent equipment failures, streamline operations, and significantly reduce energy consumption. Additionally, the use of automation and robotics increases production speed and reduces errors, while asset tracking and worker safety systems ensure optimal resource allocation and a safer work environment. The practical examples provided highlight notable time and cost savings, demonstrating IIoT's potential to revolutionize manufacturing operations and drive substantial economic benefits.

CHAPTER 4: HEALTHCARE IOT

REMOTE PATIENT MONITORING

Remote patient monitoring (RPM) is a cornerstone of Healthcare

IoT, transforming patient care by enabling continuous health

data collection and analysis outside traditional healthcare

settings. This innovative approach leverages interconnected

devices to track vital signs, medication adherence, and other

health metrics in real-time, thus facilitating timely medical

interventions and personalized treatment plans.

- **Example 1:** A diabetic patient uses a glucose monitor
  connected to their smartphone, sending frequent glucose
  level updates directly to their healthcare provider.

- **Example 2:** Heart failure patients wear smart biosensors
  to continuously monitor their heart rate and rhythm,
  alerting their physicians to any anomalies.

- **Example 3:** Seniors use wearable fitness trackers that
  monitor physical activity and sleep patterns, data that is
  shared with their caregivers and doctors.

- **Example 4:** Asthma patients utilize smart inhalers that track usage and provide data on environmental triggers.
- **Example 5:** A patient recovering from surgery uses a connected blood pressure cuff that sends readings to their medical team daily.

BENEFITS AND COST SAVINGS

1. **Improved Patient Outcomes:** Continuous monitoring helps detect potential health issues before they become serious, leading to better clinical outcomes.

2. **Reduced Hospital Readmissions:** Timely interventions can prevent complications that require hospitalization, saving up to $8,000 per readmission.

3. **Enhanced Patient Engagement:** Patients become more proactive in managing their health, reducing the need for frequent doctor's visits.

4. **Operational Efficiency:** Healthcare providers can prioritize resources and attention based on real-time data, improving overall efficiency in patient management.

5. **Cost Savings:** By minimizing the need for in-person consultations and hospital admissions, RPM can reduce healthcare costs significantly. For example, one study showed that RPM reduced total annual healthcare costs per patient by $1,500 on average.

WEARABLE HEALTH DEVICES

Wearable health devices represent a significant advancement in healthcare IoT, making continuous health monitoring more accessible and convenient for patients. These devices, which include smartwatches, fitness trackers, and biosensors, can track a variety of health metrics such as heart rate, physical activity, sleep quality, and more. By providing real-time data, healthcare providers can gain deeper insights into their patients' health and tailor interventions accordingly. This not only empowers patients to take control of their own health but also enhances the ability of medical professionals to deliver timely, accurate care.

EXAMPLES

1. **Smartwatches:** Devices like the Apple Watch can monitor heart rate, detect irregular rhythms, and send ECG readings to doctors.

2. **Fitness Trackers:** Wearables like Fitbit track daily steps, calories burned, and sleep patterns, and provide insights to both users and healthcare providers.

3. **Biosensors:** Devices that can be worn on the body to continuously monitor glucose levels for diabetics, sending alerts when intervention is needed.

4. **Smart Clothing:** Garments embedded with sensors that track muscle activity and posture, aiding in physical therapy and rehabilitation.

5. **Hearing Aids:** Advanced hearing aids that connect to smartphones, allowing users to adjust settings and track hearing health over time.

BENEFITS AND COST SAVINGS

1. **Enhanced Personalization:** Wearable devices provide detailed health data, enabling personalized treatment plans that cater to individual patient needs.

2. **Early Detection of Health Issues:** Real-time monitoring can detect potential health problems early, leading to timely interventions and reducing the risk of severe complications.

3. **Increased Patient Engagement:** Wearables encourage patients to stay engaged with their health by providing accessible metrics, which can lead to improved lifestyle and better health outcomes.

4. **Operational Efficiency:** Healthcare providers can manage patient workloads more effectively by using data from wearables to prioritize care based on urgency and need.

5. **Cost Reductions:** By minimizing the frequency of unnecessary doctor visits and hospital admissions, wearable health devices can lead to substantial cost savings. For example, proactive monitoring can save an

average of $1,200 per patient annually in healthcare costs.

TELEMEDICINE

Telemedicine, powered by IoT, has revolutionized the way healthcare services are delivered by allowing remote access to medical consultations, diagnostics, and treatment. This synergy between telemedicine and IoT enhances patient care by facilitating real-time communication and data sharing between patients and healthcare providers. Examples of telemedicine applications leveraging IoT include virtual doctor visits using video conferencing tools, remote diagnostics utilizing connected medical devices, tele-rehabilitation programs with real-time monitoring, IoT-based chronic disease management systems, and e-prescriptions facilitated through digital communication platforms. The benefits and cost savings associated with telemedicine and IoT are substantial: it increases access to healthcare services for patients in remote or underserved areas, improves the efficiency of medical practices by reducing the number of in-person visits, saves time and travel costs for patients, enables continuous monitoring that can lead to better health outcomes, and reduces overall healthcare expenditure by preventing hospital readmissions and emergency room visits.

Looking ahead, the future capabilities of telemedicine and IoT are poised to be even more transformative. As technology advances, we can expect the development of more sophisticated wearable devices that provide comprehensive health monitoring, the integration of artificial intelligence to assist in diagnosing and predicting health issues, and the enhancement of telehealth platforms with augmented reality for more interactive and effective consultations. Additionally, 5G technology will enable faster and more reliable data transmission, further improving the quality and accessibility of telehealth services. The convergence of telemedicine and IoT will continue to break down barriers to healthcare access, ensuring a more connected and efficient healthcare system.

CHAPTER SUMMARY

This chapter explores the transformative potential of IoT in healthcare, emphasizing the benefits of remote patient monitoring, wearable health devices, and telemedicine. By leveraging IoT technologies, healthcare providers can enhance patient engagement, operational efficiency, and cost savings. Wearable devices such as smartwatches, fitness trackers, and biosensors facilitate continuous health monitoring, enabling personalized care and early detection of health issues. Telemedicine, powered by IoT, revolutionizes healthcare delivery through remote consultations and real-time data sharing. Together, these advancements promise a more connected, efficient, and patient-centric healthcare system, with future innovations like artificial intelligence and 5G technology poised to further elevate the field.

CHAPTER 5: AGRICULTURAL IOT

PRECISION FARMING

In recent years, Agricultural IoT has emerged as a crucial aspect of modern farming, transforming traditional agricultural practices into technologically advanced systems. Precision farming, powered by IoT, leverages data analytics, sensors, and cloud computing to optimize farming operations, enhance crop yields, and minimize environmental impact. By integrating IoT technologies into agricultural processes, farmers can gain precise insights into soil conditions, weather patterns, and crop health, leading to more informed decisions and efficient resource use.

EXAMPLES OF AGRICULTURAL IOT APPLICATIONS

1. **Soil Moisture Sensors:** These sensors monitor soil hydration levels in real-time, ensuring optimal irrigation and preventing water wastage.

2. **Climate Monitoring Stations:** Deployed across fields, these stations gather weather data to help predict weather patterns and preemptively manage crops.

3. **Smart Irrigation Systems:** Automated irrigation systems adjust water delivery based on real-time soil moisture data, reducing water usage and increasing crop health.

4. **Drones for Crop Monitoring:** Equipped with cameras and sensors, drones capture aerial images of fields to assess crop health, detect pests, and identify areas requiring attention.

5. **Livestock Monitoring Devices:** Wearable devices on livestock collect data on health, activity levels, and location, enhancing animal welfare and farm management.

BENEFITS AND COST SAVINGS OF AGRICULTURAL IOT

1. **Increased Crop Yields:** Precision farming techniques maximize crop production by ensuring optimal growing conditions and reducing crop failure rates.

2. **Resource Efficiency:** IoT-based systems enable efficient use of water, fertilizers, and pesticides, significantly reducing waste and environmental impact.

3. **Reduced Labor Costs:** Automation of tasks such as irrigation and crop monitoring reduces the need for manual labor, leading to significant cost savings.

4. **Early Detection of Issues:** Continuous monitoring allows for early identification of pests, diseases, and nutrient deficiencies, preventing widespread crop damage and loss.

5. **Enhanced Data-Driven Decision Making:** With real-time data collection and analysis, farmers can make more accurate and timely decisions, improving overall farm productivity and sustainability.

CROP MONITORING

Crop monitoring is a pivotal aspect of precision farming, serving as the backbone for effective agricultural management. By employing a range of IoT technologies, farmers can closely observe crop growth, health, and development. This real-time monitoring enables them to react swiftly to any issues such as pest infestations, nutrient deficiencies, or water stress, ensuring that prompt corrective measures are taken. This not only minimizes crop losses but also maximizes yield quality and quantity. The data collected from crop monitoring systems provides invaluable insights that help farmers optimize inputs, reduce wastage, and enhance overall farm sustainability.

EXAMPLES OF CROP MONITORING APPLICATIONS

1. **Satellite Imagery:** Provides large-scale and detailed views of crop health and growth stages, allowing for the identification of variations within fields.

2. **Remote Sensing Technologies:** Utilize spectral imaging to detect stress symptoms in plants well before they are visible to the naked eye.

3. **Field Scanners:** Ground-based devices that scan crops to monitor growth rates and detect potential issues.

4. **Automated Sentinels:** Robotic systems that traverse fields, capturing data on plant development and identifying areas in need of intervention.

5. **Integrated Pest Management Systems:** Combine sensors and predictive analytics to forecast pest outbreaks and recommend timely actions.

BENEFITS AND COST SAVINGS OF CROP MONITORING

1. **Improved Yield Quality and Quantity:** Early detection and resolution of crop issues result in healthier plants and higher production rates.

2. **Reduction in Input Costs:** Efficient use of resources like water, fertilizers, and pesticides due to precise application based on real-time data.

3. **Lower Risk of Crop Failure:** Continuous monitoring helps preempt and address potential threats, thereby reducing the risk of total crop losses.

4. **Enhanced Sustainability:** Optimized resource use and reduced wastage lead to more environmentally friendly farming practices.

5. **Better Data-Driven Decisions:** Access to comprehensive, real-time information allows farmers to make informed decisions, boosting overall farm productivity and profitability.

LIVESTOCK TRACKING

Livestock tracking has revolutionized the way farmers manage their herds, offering a suite of technologies that provide real-time data on animal health, behavior, and location. By employing IoT devices such as GPS trackers, RFID tags, and biometric sensors, farmers can continuously monitor the well-being of their livestock. This improves not only the efficiency of herd management but also enhances the animals' overall health and productivity. Moreover, sophisticated data analytics can predict potential health issues, optimize feeding schedules, and streamline breeding programs, ultimately driving down operational costs and increasing farm profitability.

EXAMPLES OF LIVESTOCK TRACKING APPLICATIONS

1. **GPS Trackers:** Attached to animals to monitor their movements and prevent loss or theft by providing real-time location data.

2. **RFID Tags:** Used for identifying and tracking individual animals, enabling easy collection of data such as age, weight, and health records.

3. **Biometric Sensors:** Measure vital signs like heart rate and temperature to detect early signs of illness and reduce mortality rates.

4. **Smart Feeders:** Automatically dispense the right amount of feed based on individual animal needs, optimizing nutrition and growth.

5. **Wearable Health Monitors:** Devices that track physical activity, sleeping patterns, and overall health metrics to ensure optimal animal welfare.

BENEFITS AND COST SAVINGS OF LIVESTOCK TRACKING

1. **Improved Animal Health:** Early detection of illnesses allows for prompt treatment, reducing mortality rates and improving overall herd health.

2. **Enhanced Productivity:** By monitoring and optimizing feeding and breeding programs, farmers can maximize livestock growth and reproduction rates.

3. **Reduced Labor Costs:** Automation of tracking and health monitoring reduces the need for manual check-ups, saving time and labor expenses.

4. **Minimized Loss and Theft:** Real-time GPS tracking helps prevent the loss or theft of valuable livestock, ensuring better security and asset management.

5. **Better Resource Management:** Accurate data collection aids in the efficient use of resources such as feed and veterinary services, cutting down unnecessary expenditure.

IRRIGATION MANAGEMENT

Irrigation management is a critical component of modern agriculture, ensuring that crops receive the optimal amount of water needed for healthy growth and maximum yield. Advanced irrigation systems, fueled by IoT technology, allow farmers to precisely control and monitor water usage based on real-time data, weather forecasts, and soil moisture levels. This not only enhances the efficiency of water use but also reduces waste and mitigates the risks associated with both over-irrigation and under-irrigation. By investing in sophisticated irrigation management systems, farmers can achieve better crop health, increase productivity, and promote sustainable farming practices.

EXAMPLES OF IRRIGATION MANAGEMENT APPLICATIONS

1. **Smart Sprinklers:** Automated sprinkler systems that adjust water flow based on real-time weather conditions and soil moisture data.

2. **Drip Irrigation Systems:** Deliver water directly to the roots of plants, minimizing evaporation and ensuring efficient water use.

3. **Soil Moisture Sensors:** Devices placed in the ground to monitor moisture levels and provide data for optimal irrigation scheduling.

4. **Weather-Based Controllers:** Systems that use local weather data to adjust irrigation schedules, reducing water usage during rainy periods.

5. **IoT-Integrated Irrigation Software:** Platforms that offer comprehensive control and monitoring of irrigation systems through mobile and web applications.

BENEFITS AND COST SAVINGS OF IRRIGATION MANAGEMENT

1. **Water Conservation:** Efficient water usage reduces overall water consumption, lowering utility costs and conserving a vital natural resource.

2. **Enhanced Crop Yield:** Proper irrigation leads to healthier plants and higher yields, directly impacting farm profitability.

3. **Reduced Labour Costs:** Automated systems reduce the need for manual irrigation, saving time and labour expenses.

4. **Optimal Nutrient Delivery:** Controlled irrigation helps maintain soil nutrient balance, ensuring that plants receive the right amount of nutrients.

5. **Increased Sustainability:** By minimizing water waste and promoting efficient resource use, advanced irrigation management contributes to more sustainable agricultural practices.

CHAPTER SUMMARY

This chapter explores the transformative impact of IoT

technologies on modern agriculture. It highlights the benefits of

continuous crop monitoring, livestock tracking, and advanced

irrigation management. Through real-time data collection and

data-driven decision-making, these technologies enhance

productivity, sustainability, and cost-efficiency on farms. Key

applications include GPS trackers, RFID tags, biometric sensors,

smart feeders, soil moisture sensors, and weather-based

controllers. Overall, the integration of IoT in agriculture

optimizes resource management, reduces labor costs, and

promotes healthier crops and livestock.

CHAPTER 6: RETAIL IOT

INVENTORY MANAGEMENT

IoT has revolutionized inventory management in the retail sector by providing real-time visibility, accurate tracking, and efficient management of stock levels. Through IoT-enabled devices, retailers can monitor inventory in real-time, ensuring that products are always available when customers need them while avoiding overstocking. This precise control leads to better demand forecasting and inventory optimization.

EXAMPLES OF IOT APPLICATIONS IN INVENTORY MANAGEMENT

1. **RFID Tags:** These tags help track items throughout the supply chain, from warehouse to storefront, ensuring accurate inventory counts.

2. **Smart Shelves:** Equipped with weight sensors, these shelves notify staff when stock levels are low or items are misplaced.

3. **Automated Reordering:** IoT systems can automatically reorder products when inventory falls below predefined levels, preventing stockouts.

4. **Temperature Sensors:** Used for perishable goods, these sensors ensure items are stored at the correct temperature, reducing spoilage.

5. **Drones for Warehouse Management:** Drones can perform inventory checks in large warehouses, significantly speeding up the process and improving accuracy.

BENEFITS AND COST SAVINGS OF IOT IN INVENTORY MANAGEMENT

1. **Reduced Stockouts:** Real-time tracking helps maintain optimal stock levels, reducing lost sales opportunities and improving customer satisfaction.

2. **Lower Holding Costs:** Efficient inventory management reduces the need for excess stock, lowering storage costs.

3. **Minimized Shrinkage:** Enhanced tracking reduces inventory shrinkage due to theft or loss, preserving revenue.

4. **Improved Demand Forecasting:** Accurate data collection enables better prediction of demand trends, optimizing stock levels.

5. **Operational Efficiency:** Automation and accurate inventory data streamline operations, reducing labor costs and freeing up staff for more strategic tasks.

HOW IOT HAS ENHANCED PERSONALIZED SHOPPING EXPERIENCES

The integration of IoT in the retail industry has significantly transformed personalized shopping experiences by leveraging data and connectivity to tailor offerings to individual customers' preferences and behaviors. IoT devices such as beacons, smart mirrors, and connected mobile apps gather and analyze customer data, enabling retailers to provide customized recommendations, targeted promotions, and seamless in-store navigation.

EXAMPLES OF IOT APPLICATIONS IN PERSONALIZED SHOPPING EXPERIENCES

1. **Beacons:** These small devices communicate with shoppers' smartphones to deliver personalized discounts and recommendations based on their location within the store.

2. **Smart Mirrors:** Allow customers to virtually try on clothes and accessories, providing recommendations based on their selections and previous purchase history.

3. **Connected Mobile Apps:** Provide personalized content, loyalty rewards, and convenient in-store navigation, enhancing the shopping experience.

4. **Interactive Displays:** Engage customers with tailored content and information about products, helping them make informed purchasing decisions.

5. **Customer Engagement Platforms:** Analyze data from various touchpoints to create detailed customer profiles, enabling more precise targeting of marketing efforts.

BENEFITS AND COST SAVINGS OF IOT IN PERSONALIZED SHOPPING EXPERIENCES

1. **Enhanced Customer Satisfaction:** Personalized shopping experiences increase customer satisfaction and loyalty, driving repeat business.

2. **Increased Sales:** Tailored recommendations and targeted promotions boost sales by connecting customers with products relevant to their interests.

3. **Better Inventory Management:** Understanding customer preferences helps retailers stock more relevant products, reducing waste and overstocking.

4. **Reduced Marketing Costs:** Targeted advertising ensures marketing budgets are spent more effectively, reaching the right audience with the right message.

5. **Operational Efficiency:** Automating customer data collection and analysis frees up staff to focus on providing high-quality service and strategic tasks.

HOW IOT HAS HELPED WITH BEACONS AND PROXIMITY MARKETING

The incorporation of IoT technologies, particularly beacons and proximity marketing, has revolutionized how retailers interact with consumers by delivering highly targeted and contextualized content. Beacons, small wireless transmitters, use Bluetooth technology to send signals to nearby smart devices, allowing retailers to engage customers with personalized offers and information based on their precise location within a store. This proximity-driven approach not only enhances the shopping experience but also drives foot traffic and boosts sales.

EXAMPLES OF IOT APPLICATIONS IN BEACONS AND PROXIMITY MARKETING

1. **In-Store Notifications:** Beacons send push notifications to customers' smartphones about ongoing sales and exclusive offers when they enter specific store zones.

2. **Personalized Greetings:** Upon entering a store, registered customers receive a personalized greeting and tailored recommendations based on their purchase history.

3. **Location-Based Ads:** Retailers provide location-specific advertisements and promotions, increasing the relevance and effectiveness of marketing campaigns.

4. **Interactive Store Maps:** Beacons help customers navigate large stores by providing interactive maps and directions to desired products or sections.

5. **Analytical Insights:** Data collected from beacon interactions informs retailers about customer behavior patterns, enhancing marketing strategies and store layouts.

BENEFITS AND COST SAVINGS OF IOT IN BEACONS AND PROXIMITY MARKETING

1. **Increased Customer Engagement:** Personalized content and seamless navigation keep customers engaged and more likely to explore the store and make purchases.

2. **Boosted Sales:** Targeted promotions and recommendations drive impulse buys and improve overall sales figures.

3. **Enhanced Loyalty Programs:** Tailored rewards and offers strengthen customer loyalty, leading to repeat business and long-term relationships.

4. **Optimized Marketing Spend:** Precise targeting reduces wasted marketing efforts and ensures that promotional budgets are used more effectively.

5. **Operational Efficiency:** Automating customer interactions and data collection allows staff to focus on more strategic tasks and improve in-store service.

HOW IOT HAS HELPED WITH SUPPLY CHAIN OPTIMIZATION

The integration of IoT in supply chain management has revolutionized how businesses monitor, track, and manage their supply chain operations. By leveraging IoT devices such as sensors, RFID tags, and GPS trackers, companies gain real-time visibility into the movement and condition of goods throughout the supply chain. This enhanced transparency enables better decision-making, reducing inefficiencies and optimizing overall operations.

EXAMPLES OF IOT APPLICATIONS IN SUPPLY CHAIN OPTIMIZATION

1. **Real-Time Tracking:** GPS-equipped IoT devices provide live updates on the location of shipments, helping companies maintain accurate delivery schedules.

2. **Condition Monitoring:** Sensors monitor temperature, humidity, and other environmental conditions to ensure that perishable goods are transported under appropriate conditions.

3. **Automated Inventory Management:** RFID tags and smart shelves automatically track inventory levels, reducing manual errors and streamlining stock replenishment.

4. **Predictive Maintenance:** IoT sensors on machinery predict maintenance needs before they result in breakdowns, minimizing downtime and repair costs.

5. **Fleet Management:** IoT solutions track vehicle metrics such as fuel consumption and engine health, optimizing route planning and improving fleet efficiency.

BENEFITS AND COST SAVINGS OF IOT IN SUPPLY CHAIN OPTIMIZATION

1. **Improved Efficiency:** Real-time data and analytics streamline supply chain processes, reducing delays and ensuring smooth operations.

2. **Cost Reduction:** Enhanced inventory management and predictive maintenance lower operational and maintenance costs.

3. **Better Decision-Making:** Accurate and timely data enable more informed decision-making across the supply chain, reducing risks and inefficiencies.

4. **Enhanced Customer Satisfaction:** Real-time tracking and condition monitoring ensure timely and safe delivery of products, boosting customer satisfaction.

5. **Operational Visibility:** Increased transparency across the supply chain allows for greater control and agility, adapting quickly to changes in demand or disruptions.

CHAPTER SUMMARY

In this chapter, we explored the transformative impact of IoT technologies on the retail and supply chain sectors. We discussed how IoT, through devices like beacons and sensors, enables personalized customer engagement, enhances inventory management, optimizes marketing efforts, and improves operational efficiency in retail environments. Additionally, we examined the role of IoT in supply chain management, highlighting its benefits in real-time tracking, condition monitoring, automated inventory management, predictive maintenance, and fleet management. The integration of IoT across these domains not only drives customer satisfaction and loyalty but also ensures cost savings and streamlined operations, ultimately leading to more agile and resilient business practices.

CHAPTER 7: ENERGY MANAGEMENT IOT

SMART GRID TECHNOLOGY

Smart grid technology represents a substantial advancement in how we manage and distribute electricity. By integrating IoT devices, utilities can monitor and control energy flow more efficiently, improving the reliability and sustainability of power distribution. Smart grids facilitate real-time communication between utility providers and customers, allowing for better demand response and energy usage optimization. This intelligent network not only helps in reducing energy waste but also enhances the resilience of the electrical grid against outages and peak demands. The benefits of smart grid technology extend to both consumers and providers, delivering substantial cost savings and operational efficiency.

EXAMPLES OF SMART GRID IOT APPLICATIONS

1. **Smart Meters:**

- **Benefit:** Provide real-time data on electricity usage to both utility companies and consumers.

- **Cost Savings:** Reduce the need for manual meter readings, lowering operational costs.

2. **Demand Response Systems:**

- **Benefit:** Automatically adjust energy consumption based on grid load and pricing signals.

- **Cost Savings:** Minimize peak demand charges and reduce energy expenditure during high-cost periods.

3. **Distributed Energy Resources (DER) Integration:**

- **Benefit:** Allows for the incorporation of renewable energy sources like solar and wind into the grid.

- **Cost Savings:** Decrease dependency on fossil fuels and lower energy costs through the use of locally generated renewable energy.

4. **Advanced Distribution Management Systems (ADMS):**

- **Benefit:** Enhance grid reliability and efficiency by providing real-time monitoring and control.

- **Cost Savings:** Reduce outage times and maintenance costs by promptly addressing issues within the grid.

5. **Electric Vehicle (EV) Charging Management:**

- **Benefit:** Optimize EV charging schedules to ensure grid stability and efficient energy use.

- **Cost Savings:** Avoid overloading the grid and lower operational costs by balancing load during off-peak hours.

ENERGY MONITORING AND OPTIMIZATION

IoT technology significantly enhances energy monitoring and optimization by providing detailed insights into energy usage patterns and enabling automated control of various systems. This data-driven approach allows both consumers and providers to optimize energy consumption, reduce waste, and lower costs.

EXAMPLES OF IOT APPLICATIONS IN ENERGY MONITORING AND OPTIMIZATION

1. **Smart Thermostats:**

- **Benefit: Automatically adjust heating and cooling based on occupancy and weather conditions.**

- Cost Savings: **Lower energy bills by optimizing HVAC usage and reducing unnecessary energy consumption.**

2. Automated Lighting Systems:

- Benefit: **Control and schedule lighting based on occupancy and natural light availability.**

- Cost Savings: **Significant reduction in electricity usage by minimizing unnecessary lighting and optimizing usage during peak and off-peak hours.**

3. Energy Management Systems (EMS):

- Benefit: **Provide real-time data on energy consumption across facilities, enabling detailed analysis and optimization.**

- Cost Savings: **Reduce overall energy expenditures by identifying inefficiencies and implementing energy-saving measures.**

4. Predictive Maintenance for Equipment:

- Benefit: **Monitor the performance of machinery and predict maintenance needs before breakdowns occur.**

- Cost Savings: **Decrease maintenance costs and downtime by addressing issues proactively rather than reactively.**

5. Home Energy Management Systems (HEMS):

- Benefit: **Enable consumers to monitor and control home energy usage through mobile apps and smart devices.**

- Cost Savings: **Empower users to make informed decisions about their energy usage, leading to lower utility bills and better energy management.**

DEMAND RESPONSE SYSTEMS

IoT plays a pivotal role in demand response systems by enabling real-time communication and automated energy management based on grid conditions and pricing signals. By leveraging IoT technology, these systems can dynamically adjust energy consumption patterns, which helps in balancing supply and demand, reducing peak loads, and ensuring grid stability. This not only enhances grid efficiency but also results in significant cost savings for both utility providers and consumers.

EXAMPLES OF IOT APPLICATIONS IN DEMAND RESPONSE SYSTEMS

1. **Smart Appliances:**

- **Benefit:** Automatically adjust operation schedules based on energy availability and pricing.

- **Cost Savings:** Reduce energy bills by shifting usage to off-peak times when electricity is cheaper.

1. **Automated HVAC Systems:**

- **Benefit:** Control heating, ventilation, and air conditioning to align with demand response signals.

- **Cost Savings:** Minimize operational costs by lowering energy usage during peak periods and improving overall energy efficiency.

1. **Smart Plug Load Controllers:**

- **Benefit:** Manage and control plug-in devices to reduce unnecessary energy consumption.

- **Cost Savings:** Cut down on standby power and lower electricity costs by turning off devices when not in use or during peak pricing periods.

1. **Battery Storage Systems:**

- **Benefit:** Store excess energy during low-demand periods and discharge stored energy during high-demand times.

- **Cost Savings:** Avoid peak energy charges and improve energy reliability through self-sufficient energy storage solutions.

1. **Dynamic Pricing Programs:**

- **Benefit:** Provide consumers with real-time pricing information to encourage energy usage shifts.

- **Cost Savings:** Enable consumers to make informed decisions about their energy consumption, leading to reduced utility bills by taking advantage of lower energy rates during off-peak periods.

RENEWABLE ENERGY INTEGRATION

IoT facilitates the integration of renewable energy sources into the grid by providing advanced monitoring, control, and management capabilities. These technologies enhance the efficiency and reliability of renewable energy systems, ensuring optimal performance and seamless incorporation into existing infrastructure. By enabling real-time data collection and analysis, IoT solutions can forecast energy production, optimize storage, and manage distribution, ultimately leading to better utilization of renewable resources and reduced dependency on fossil fuels.

EXAMPLES OF IOT APPLICATIONS IN RENEWABLE ENERGY INTEGRATION

1. **Smart Solar Inverters:**

- **Benefit:** Optimize the conversion of solar energy into usable electricity.

- **Cost Savings:** Increase the efficiency of solar panels, leading to higher energy production and lower overall costs.

2. **Wind Turbine Monitoring Systems:**

- **Benefit:** Continuously monitor wind turbine performance and weather conditions.

- **Cost Savings:** Enhance maintenance schedules and reduce downtime by addressing issues before they escalate.

3. **Energy Storage Management Systems:**

- **Benefit:** Efficiently manage the storage and release of energy from batteries.

- **Cost Savings:** Maximize the use of stored energy, reducing reliance on the grid during peak times and lowering energy costs.

4. **Microgrid Controllers:**

- **Benefit:** Coordinate the operation of a network of distributed energy resources.

- **Cost Savings:** Improve energy resilience and lower operational costs by optimizing the use of local renewable energy sources.

5. **Real-Time Grid Monitoring:**

- **Benefit:** Provide detailed insights into grid performance and energy flows.

- **Cost Savings:** Enhance grid stability and efficiency, reducing the need for expensive infrastructure upgrades and improving the integration of renewables.

CHAPTER SUMMARY

This chapter explores the significant role of IoT in enhancing energy efficiency and integrating renewable energy. Key topics include the benefits and cost savings of automated lighting systems, energy management systems, predictive maintenance, and home energy management systems. Additionally, it delves into demand response systems, highlighting examples like smart appliances and automated HVAC systems. Lastly, the chapter covers renewable energy integration via IoT applications such as smart solar inverters, wind turbine monitoring, and microgrid controllers, emphasizing the improved efficiency and reduced costs facilitated by these technologies.

CHAPTER 8: TRANSPORTATION IOT

CONNECTED VEHICLES

The advent of IoT in the transportation sector, specifically connected vehicles, is revolutionizing how we interact with and manage transportation systems. Connected vehicles utilize IoT technologies to improve safety, enhance operational efficiency, and reduce costs. By fostering real-time communication between vehicles and infrastructure, IoT enables advanced traffic management, predictive maintenance, and enhanced navigation systems. These innovations contribute to a safer and more efficient transportation ecosystem, ultimately leading to significant cost savings for both consumers and businesses. Below are five examples of IoT applications in connected vehicles, along with their benefits and cost savings.

1. **Vehicle-to-Everything (V2X) Communication:**
- **Benefit:** Enhances road safety by enabling vehicles to communicate with each other and with traffic infrastructure.

- **Cost Savings:** Reduces accident-related costs by preventing collisions and optimizing traffic flow, leading to lower fuel consumption.

2. **Predictive Maintenance Systems:**

- **Benefit:** Uses real-time data to forecast potential vehicle issues before they cause breakdowns.

- **Cost Savings:** Lowers maintenance costs by scheduling repairs proactively, reducing the likelihood of costly emergency repairs.

3. **Driver Behavior Monitoring:**

- **Benefit:** Tracks and analyzes driver behavior to improve driving habits and enhance safety.

- **Cost Savings:** Decreases insurance premiums by reducing risky driving incidents and promoting safer driving practices.

4. **Fleet Management Solutions:**

- **Benefit:** Allows companies to monitor and manage their vehicle fleets in real-time.

- **Cost Savings:** Optimizes routes, reduces fuel consumption, and improves vehicle utilization, leading to substantial operational cost reductions.

5. **Autonomous Driving Features:**

- **Benefit:** Enhances driving assistance through advanced sensors and real-time data processing.

- **Cost Savings:** Reduces the need for human drivers in commercial fleets, lowering labor costs and minimizing human error-related expenses.

FLEET MANAGEMENT

The integration of IoT in fleet management has profoundly improved operational efficiency and cost-effectiveness. By leveraging IoT technologies, fleet managers can gain real-time insights into vehicle locations, performance, and maintenance needs. This data-driven approach enables proactive decision-making, reduces fuel consumption, and minimizes vehicle downtime. IoT also enhances route optimization, driver safety, and compliance with regulatory requirements, ultimately leading to a highly streamlined and profitable fleet operation. Below are five examples of IoT applications in fleet management, along with their costs and benefits.

1. **Telematics Systems:**

- **Benefit:** Provides comprehensive data on vehicle performance, driver behavior, and fuel usage.

- **Cost Savings:** Reduces fuel costs by up to 15% through optimized driving.

2. **Asset Tracking:**

- **Benefit:** Monitors the real-time location and status of assets and cargo.

- **Cost Savings:** Decreases asset loss and theft-related costs by ensuring better security and accountability.

3. **Predictive Maintenance:**

- **Benefit:** Predicts and addresses maintenance issues before they result in breakdowns.

- **Cost Savings:** Cuts maintenance costs by up to 30% by preventing costly repairs and extending vehicle life.

4. **Route Optimization Software:**

- **Benefit:** Analyzes traffic and road conditions to determine the most efficient routes.

- **Cost Savings:** Reduces fuel consumption and delivery times, lowering operational expenses by approximately 20%.

5. **Fleet Safety Solutions:**

- **Benefit:** Implements safety measures like real-time alerts and driver support systems.

- **Cost Savings:** Lowers accident-related costs and insurance premiums by up to 25% through enhanced safety protocols.

HOW IOT IMPROVES TRAFFIC MANAGEMENT

The integration of IoT in traffic management systems offers transformative benefits by enabling real-time data collection and analysis. This technology enhances traffic flow, reduces congestion, and improves overall road safety. With IoT, traffic signals, cameras, and sensors can communicate seamlessly to provide dynamic traffic control, data-driven decision-making, and timely interventions. These advancements result in a well-coordinated traffic network that minimizes delays, reduces fuel consumption, and lowers emission levels, ultimately creating a more efficient and sustainable urban environment. Below are five examples of IoT applications in traffic management, along with their costs and benefits.

1. **Adaptive Traffic Signal Systems:**

- **Benefit:** Adjusts traffic light timings based on real-time traffic conditions.

- **Cost Savings:** Reduces waiting time at intersections by up to 30%, leading to lower fuel consumption and emissions.

2. **Smart Parking Solutions:**

- **Benefit:** Guides drivers to available parking spaces using real-time data.

- **Cost Savings:** Decreases search time for parking by up to 50%, reducing fuel use and traffic congestion.

3. **Traffic Flow Monitoring:**

- **Benefit:** Collects and analyzes traffic data to optimize road usage.

- **Cost Savings:** Improves road usage efficiency, reducing traffic jams and associated economic costs.

4. **Incident Detection and Response Systems:**

- **Benefit:** Identifies and responds to accidents or road hazards promptly.

- **Cost Savings:** Minimizes response time, decreasing secondary accidents and traffic delays, leading to fewer economic losses from traffic disruptions.

5. **Public Transportation Management:**

- **Benefit:** Integrates IoT to provide real-time updates and optimize routes for public transport.

- **Cost Savings:** Enhances service efficiency, leading to increased ridership and reduced operational costs for transit authorities.

AUTONOMOUS VEHICLES

The implementation of IoT in autonomous vehicles significantly enhances their performance, reliability, and safety. Through real-time data collection and analysis, IoT enables autonomous vehicles to make precise and informed decisions that improve navigation, obstacle detection, and adaptive driving. This connectivity ensures that autonomous vehicles can communicate seamlessly with each other and with infrastructure, leading to better route planning and traffic management. Consequently, IoT integration not only boosts the efficiency of autonomous vehicles but also reduces operational costs and increases driving safety. Here are five examples of IoT applications in autonomous vehicles, along with their costs and benefits:

1. **Real-Time Mapping:**

Benefit: Provides up-to-date maps and traffic information for precise navigation.

Cost Savings: Reduces fuel consumption and travel time by avoiding traffic congestion and taking optimal routes.

2. **Obstacle Detection Systems:**

Benefit: Uses sensors and cameras to detect and respond to obstacles on the road.

Cost Savings: Minimizes accident-related costs by preventing collisions and reducing insurance premiums.

3. **Vehicle-to-Infrastructure Communication:**

Benefit: Enables vehicles to receive real-time information from traffic signals and road infrastructure.

Cost Savings: Decreases delays at intersections, leading to lower fuel use and emissions.

4. **Remote Diagnostics:**

Benefit: Continuously monitors vehicle health and performance, alerting to potential issues.

Cost Savings: Lowers maintenance expenses by addressing problems early and avoiding costly repairs.

5. **Platooning Technology:**

Benefit: Allows autonomous vehicles to travel in tightly coordinated groups, reducing drag.

Cost Savings: Enhances fuel efficiency by up to 15%, resulting in significant fuel cost reductions.

CHAPTER SUMMARY

This chapter delves into the transformative impact of IoT on various aspects of vehicle and traffic management. It illustrates how IoT technologies enhance fleet management by providing comprehensive data, optimizing routes, and predicting maintenance needs, which collectively reduce operational costs and improve safety. Furthermore, the chapter highlights the role of IoT in traffic management, demonstrating how adaptive systems and real-time monitoring can alleviate congestion and lower emissions. Finally, the chapter underscores the importance of IoT in the evolution of autonomous vehicles, showcasing its role in improving navigation, obstacle detection, and vehicle-to-infrastructure communication, thereby increasing efficiency and reducing costs.

CHAPTER 9: WEARABLE TECHNOLOGY

FITNESS TRACKERS

Fitness trackers have revolutionized the way individuals monitor their health and wellness routines. By providing real-time data on various metrics such as steps taken, heart rate, and sleep patterns, these devices empower users to make informed decisions about their lifestyle. The integration of IoT in fitness trackers enables seamless connectivity with smartphones and other devices, allowing for comprehensive tracking and analysis of fitness goals. Additionally, the ease of use and portability of fitness trackers make them an essential tool for anyone looking to maintain or improve their physical health.

Here are five examples of fitness tracker applications, along with their benefits:

1. **Step Counting:**

- **Benefit:** Tracks the number of steps taken daily, encouraging users to meet their activity goals.

- **Motivation:** Provides reminders and rewards for achieving step targets, promoting an active lifestyle.

2. **Heart Rate Monitoring:**

- **Benefit:** Continuously measures heart rate during rest and exercise.

- **Health Insight:** Alerts users to abnormal heart rates, enabling timely medical consultations.

3. **Sleep Tracking:**

- **Benefit:** Analyzes sleep duration and quality.

- **Well-being:** Helps users understand their sleep patterns and make adjustments for better rest.

4. **Calorie Tracking:**

- **Benefit:** Estimates calories burned based on activity levels and provides dietary recommendations.

- **Weight Management:** Aids users in managing their calorie intake and maintaining a healthy weight.

5. **Goal Setting and Reminders:**

- **Benefit:** Allows users to set personalized fitness goals and receive reminders.

- **Achievement Tracking:** Helps users stay on track and measure progress towards their fitness milestones.

SMART CLOTHING

Smart clothing, also known as e-textiles or wearable technology, represents an innovative fusion of fashion and functionality. These garments embed sensors and electronic components directly into fabrics, enabling the monitoring and collection of a wide range of physiological data. With applications ranging from health monitoring to enhanced athletic performance, smart clothing offers a discreet and comfortable way to gather real-time information, facilitating proactive health management and optimized training routines. The integration of IoT in smart clothing provides seamless data synchronization with smartphones and other devices, ensuring that users have access to valuable insights at their fingertips.

Here are five examples of smart clothing applications, along with their benefits:

1. **Heart Rate Monitoring Shirts:**

Benefit: Continuously tracks heart rate during exercise and daily activities.

Health Monitoring: Helps users maintain optimal training intensity and monitor cardiovascular health.

2. **Temperature-Regulating Jackets:**

Benefit: Adjusts internal temperature based on external conditions.

Comfort: Keeps users comfortable in varying weather conditions by maintaining a consistent temperature.

3. **Posture-Correcting Shirts:**

Benefit: Provides real-time feedback on posture.

Postural Health: Encourages better posture habits, reducing the risk of strain and injury.

4. **GPS-Enabled Clothing:**

Benefit: Incorporates GPS technology for tracking location.

Safety: Ideal for outdoor enthusiasts, ensuring they can navigate and be located easily in case of emergencies.

5. **Muscle Recovery Garments:**

Benefit: Utilizes compression and embedded sensors to monitor muscle activity and recovery.

Performance Enhancement: Supports faster recovery times and reduces muscle soreness after intense workouts.

HEALTH MONITORING DEVICES

Health monitoring devices have become an integral part of personal health management, offering continuous tracking and real-time feedback on various health metrics. These devices range from simple wearable gadgets to sophisticated clinical-grade equipment, providing invaluable data that can aid in the early detection of health issues and the management of chronic conditions. By integrating advanced sensors and IoT technologies, health monitoring devices enable users to monitor vital signs, track activity levels, and even manage medication schedules, all through a connected ecosystem.

Here are five examples of health monitoring devices, along with their benefits:

1. **Blood Pressure Monitors:**

- **Benefit:** Measures and tracks blood pressure throughout the day.

- **Health Management:** Helps users monitor hypertension and manage cardiovascular health.

2. **Glucose Monitors:**

- **Benefit:** Continuously tracks blood sugar levels.

- **Diabetes Management:** Allows individuals with diabetes to maintain optimal blood glucose levels and adjust their medication accordingly.

3. **Wearable ECG Monitors:**

- **Benefit:** Records electrocardiogram data in real time.

- **Heart Health:** Enables early detection of arrhythmias and other heart conditions, allowing for prompt medical intervention.

4. **Smart Thermometers:**

- **Benefit:** Provides accurate and instant body temperature readings.

- **Fever Management:** Assists in the monitoring of fever and tracking temperature fluctuations over time.

5. **Medication Reminders:**

- **Benefit:** Alerts users to take their medications on schedule.

- **Adherence Management:** Ensures that individuals do not miss doses, improving treatment outcomes and overall health.

AUGMENTED REALITY GLASSES

Augmented reality (AR) glasses are revolutionizing the way we interact with the world by overlaying digital information onto the physical environment. These advanced wearables enable users to experience a blend of virtual and real-world elements, enhancing the way we perceive and engage with various tasks, whether in professional settings, educational environments, or daily activities. By incorporating AR technology, users can access real-time data, receive navigational assistance, and even participate in immersive gaming experiences, all through a lightweight and comfortable eyewear form factor.

Here are five examples of augmented reality glasses applications, along with their benefits:

1. **Navigation Assistance:**

Benefit: Provides real-time directions and maps directly in the user's line of sight.

Convenience: Makes travel easier by offering hands-free navigation, ideal for walking, driving, or cycling.

2. **Medical Surgery Aid:**

Benefit: Overlays critical patient data and virtual anatomical guides during surgery.

Precision: Enhances surgical accuracy and efficiency by providing surgeons with vital information without taking their eyes off the patient.

3. Industrial and Technical Training:

Benefit: Delivers step-by-step instructions and visual aids for complex tasks.

Productivity: Improves the training process for technicians and workers, reducing errors and increasing understanding.

4. Remote Assistance:

Benefit: Allows experts to see what the user sees and provide live guidance.

Support: Facilitates real-time troubleshooting and support for field technicians or customer service.

5. Educational Augmentation:

Benefit: Creates interactive and engaging learning experiences by overlaying educational content in the real world.

Learning Enhancement: Makes subjects like science, history, and art more tangible and dynamic, improving student engagement and comprehension.

COSTS OF SMART CLOTHING, HEALTH MONITORING DEVICES, AND AUGMENTED REALITY GLASSES

Understanding the costs associated with these advanced technologies is crucial for making informed purchasing decisions. Here, we break down the costs of smart clothing, health monitoring devices, and augmented reality glasses, numbering them in order of their mention:

1. **Smart Clothing:**

- **Heart Rate Monitoring Shirts:** Typically range from $100 to $200, depending on the brand and additional features.

- **Temperature-Regulating Jackets:** Cost around $150 to $300, influenced by the materials used and the sophistication of the temperature control technology.

- **Posture-Correcting Shirts:** Priced between $50 and $150, varying based on the degree of feedback and sensor integration.

- **GPS-Enabled Clothing:** Can cost from $200 to $400, mainly due to the advanced GPS technology and the durability required for outdoor use.

- **Muscle Recovery Garments:** Generally range from $75 to $250, with prices reflecting the quality of compression materials and embedded sensors.

2. **Health Monitoring Devices:**

- **Blood Pressure Monitors:** Range from $40 to $100, with higher-end models offering more features and better accuracy.

- **Glucose Monitors:** Typically cost between $20 and $60, but continuous glucose monitors (CGMs) can be more expensive, ranging from $300 to $500.

- **Wearable ECG Monitors:** Prices vary widely from $80 to $300, depending on the accuracy and real-time monitoring capabilities.

- **Smart Thermometers:** Usually priced between $25 and $80, with advanced features such as Bluetooth connectivity influencing the cost.

- **Medication Reminders:** Range from $20 to $70, with more sophisticated models including smartphone integration costing on the higher end.

3. **Augmented Reality Glasses:**

- **Navigation Assistance:** AR glasses with navigation capabilities can range from $400 to $1,200, depending on the sophistication of the software and hardware.

- **Medical Surgery Aid:** Professional-grade AR glasses for surgical use can be quite expensive, ranging from $2,000 to $5,000, reflecting their advanced medical technology.

- **Industrial and Technical Training:** Typically cost between $1,500 and $3,000, with prices influenced by the complexity of the instructional software and robustness of the hardware.

- **Remote Assistance:** AR glasses designed for remote assistance usually range from $800 to $2,500, based on real-time communication features and durability for field use.

- **Educational Augmentation:** Priced between $300 and $800, influenced by the level of detail and interactivity of the educational content.

By understanding these cost ranges, consumers and professionals can better evaluate which products fit their needs and budgets, ensuring they get the most value in their investment in smart clothing, health monitoring devices, and augmented reality glasses.

CHAPTER SUMMARY

This chapter delves into the innovative world of smart clothing, health monitoring devices, and augmented reality glasses, highlighting their numerous applications and benefits. It explores various examples in each category, such as navigation assistance, medical surgery aids, and educational augmentation, showcasing how these technologies enhance daily life and professional tasks. Additionally, the chapter provides a detailed cost analysis for each type of device, offering a comprehensive guide for consumers to make informed purchasing decisions based on their needs and budgets.

CHAPTER 10: ENVIRONMENTAL MONITORING IOT

AIR QUALITY SENSORS

Environmental monitoring through the Internet of Things (IoT) involves the use of sensors and connected devices to collect real-time data on various environmental parameters. This technology plays a crucial role in ensuring environmental safety, enhancing public health, and guiding policy decisions. By continuously monitoring factors such as air and water quality, soil conditions, and noise levels, IoT devices help in mitigating environmental risks and promoting sustainable practices.

Here are five examples of environmental monitoring IoT applications, along with their costs and benefits:

1. **Air Quality Sensors:**

Cost: Typically range from $100 to $500, depending on the sensitivity and type of pollutants monitored.

Benefit: Continuously measure pollutants such as PM2.5, PM10, CO_2, and volatile organic compounds (VOCs), helping to improve public health by providing early warnings about poor air quality and guiding pollution control measures.

2. **Water Quality Monitors:**

Cost: Generally priced between $200 and $1,000, influenced by the number of parameters tested (e.g., pH, turbidity, contaminants).

Benefit: Detects contaminants and pollutants in water bodies, ensuring water safety and promoting efficient water management practices by alerting authorities to pollution events.

3. **Soil Moisture Sensors:**

Cost: Varies from $50 to $300, based on sensor accuracy and connectivity features.

Benefit: Helps in precision agriculture by providing real-time data on soil moisture levels, enabling farmers to optimize irrigation schedules and improve crop yields while conserving water.

4. **Noise Monitoring Systems:**

Cost: Typically ranges from $150 to $800, depending on the sensitivity and range of frequencies detected.

Benefit: Monitors noise pollution levels in urban areas, construction sites, and industrial zones, contributing to better noise regulation and ensuring compliance with environmental noise standards to protect public health.

5. **Weather Stations:**

Cost: Generally, between $300 and $1,500, depending on the range of meteorological parameters measured (e.g., temperature, humidity, wind speed, precipitation).

Benefit: Provides accurate and real-time weather data, crucial for disaster preparedness, climate research, and daily weather forecasting, enabling communities to better respond to weather-related hazards

WATER QUALITY MONITORING

Monitoring water quality using IoT technology enables the detailed and constant assessment of water sources, ensuring safety and sustainability. This technology plays a vital role in identifying contaminants and managing water resources effectively. By providing real-time data, water quality monitoring systems help in preventing waterborne diseases, ensuring safe drinking water, and maintaining healthy aquatic ecosystems. Here are five examples of water quality monitoring applications, along with their costs and benefits:

1. **pH Level Monitoring:**

Cost: Usually ranges from $50 to $300, based on accuracy and connectivity features.

Benefit: Ensures the water's acidity/alkalinity is within safe and optimal ranges, which is essential for both drinking water and aquatic life health.

2. **Turbidity Sensors:**

Cost: Typically priced between $100 and $600, varying with sensitivity and data resolution.

Benefit: Measures the clarity of water, which helps in detecting the presence of suspended solids, crucial for water treatment processes and environmental monitoring.

3. **Contaminant Detection Systems:**

Cost: Generally ranges from $500 to $2,000, depending on the range and specificity of contaminants detected.

Benefit: Identifies harmful contaminants like heavy metals and chemicals, ensuring the safety of drinking water and adherence to health regulations.

4. **Dissolved Oxygen Meters:**

Cost: Usually between $150 and $800, influenced by measurement range and sensor calibration quality.

Benefit: Measures the amount of dissolved oxygen in water, which is critical for maintaining aquatic life and assessing water ecosystem health.

5. **Nutrient Sensors (Nitrate/Phosphate Sensors):**

Cost: Typically ranges from $200 to $1,200, based on sensor accuracy and the number of detected nutrients.

Benefit: Monitors nutrient levels in water bodies, helping to prevent eutrophication and promoting balanced ecosystems by guiding agricultural runoff management.

WASTE MANAGEMENT

Efficient waste management is increasingly critical in our rapidly urbanizing world, and IoT technology offers innovative solutions to optimize waste collection, reduce costs, and minimize environmental impact. By deploying smart waste management systems, municipalities and businesses can monitor waste levels in real-time, streamline collection routes, and ensure timely disposal. These systems not only enhance operational efficiency but also contribute to cleaner and more sustainable urban environments.

Here are five examples of IoT applications in waste management, along with their costs and benefits:

1. **Smart Trash Bins:**

Cost: Typically range from $200 to $800, depending on features such as fill level sensors, compactors, and connectivity options.

Benefit: Monitors waste levels and alerts collection services when bins are full, reducing overflow, optimizing collection schedules, and cutting down on unnecessary collections.

2. **Fleet Management Systems:**

Cost: Usually between $1,000 and $10,000, depending on fleet size and system complexity.

Benefit: Uses GPS and sensor data to optimize waste collection routes, reducing fuel consumption, operational costs, and carbon emissions.

3. **Recycling Monitoring Systems:**

Cost: Generally priced from $500 to $2,000, varying with the type of materials monitored and system sophistication.

Benefit: Tracks recycling bins' contents to ensure proper sorting by users and improve recycling rates, leading to more efficient waste processing and reduced landfill usage.

4. **Landfill Monitoring Sensors:**

Cost: Typically ranges from $1,000 to $5,000, based on sensor types and monitoring breadth.

Benefit: **Monitors conditions such as gas emissions and leachate levels, helping manage environmental impact, ensuring regulatory compliance, and enhancing safety.**

5. Smart Waste Compactors:

Cost: **Usually between $2,000 and $10,000, influenced by size and automation level.**

Benefit: **Compacts waste on-site to increase bin capacity, reducing frequency of collections, lowering transportation costs, and minimizing environmental footprint.**

WILDLIFE TRACKING

Tracking wildlife using IoT technology has revolutionized the way scientists and conservationists monitor animal populations and their habitats. These advanced systems provide crucial insights into animal behaviors, migration patterns, and environmental threats, which can help in developing effective conservation strategies. By harnessing real-time data, wildlife tracking can mitigate human-wildlife conflicts, safeguard endangered species, and ensure the preservation of ecosystems. Below are five examples of IoT applications in wildlife tracking, along with their costs and benefits:

1. **GPS Collars:**

Cost: Typically range from $300 to $2,000, based on battery life, ruggedness, and data transmission capabilities.

Benefit: Provides precise location data, enabling the study of movement patterns and habitat use, and helping to identify critical habitats and migration corridors.

2. **Camera Traps:**

Cost: Generally between $100 and $800, depending on image resolution, trigger sensitivity, and connectivity options.

Benefit: Captures images and videos of wildlife, offering insights into animal behaviours and population densities without human interference.

3. **Acoustic Sensors:**

Cost: Usually ranges from $200 to $1,000, influenced by detection range and data processing capabilities.

Benefit: Records animal calls and sounds, useful for monitoring species presence, population estimates, and behavioral studies, particularly for elusive or nocturnal species.

4. Drone Surveillance:

Cost: Typically between $1,000 and $10,000, depending on flight range, camera quality, and autonomous capabilities.

Benefit: Provides aerial monitoring of wildlife and habitats, enabling large-area surveys and hard-to-reach area observations, reducing the need for intrusive, on-ground tracking.

5. Radio Frequency Identification (RFID) Tags:

Cost: Generally, ranges from $50 to $300 per tag, depending on size and data storage capacity.

Benefit: Offers a less invasive method for tracking smaller animals or those in sensitive environments, allowing for species-specific data collection and individual identification.

SUMMARY

This chapter highlights the transformative impact of Internet of Things (IoT) technology in both waste management and wildlife tracking. It details various IoT applications, including smart trash bins, fleet management systems, and recycling monitoring systems, which collectively enhance the efficiency of waste collection, reduce environmental impact, and lower operational costs. Additionally, the chapter explores how IoT innovations like GPS collars, camera traps, and drone surveillance are revolutionizing wildlife conservation. These technologies provide invaluable data on animal behaviours, migration patterns, and habitat use, aiding in the development of effective conservation strategies. The outlined costs and benefits associated with each application underscore the practical and financial considerations crucial for successful implementation.

CHAPTER 11: SMART CITIES AND INFRASTRUCTURE

The development of smart cities represents a significant leap towards the integration of advanced technology solutions to manage urban infrastructure efficiently. Smart cities utilize various IoT applications and data analytics to enhance the quality and performance of urban services, reduce costs, and promote sustainable development. For instance, smart lighting systems deploy sensors to optimize energy consumption based on real-time conditions, thereby lowering electricity use and reducing maintenance needs. Similarly, IoT-enabled water management systems can monitor consumption patterns and detect leaks to ensure efficient water distribution and conservation. Traffic management is another area greatly benefiting from smart city technologies; intelligent traffic systems analyze traffic flow and adjust traffic signals dynamically to minimize congestion and reduce emissions. Smart cities are also fostering greater connectivity and inclusivity by facilitating citizen engagement and public safety. IoT platforms allow residents to interact with city services through mobile apps, providing real-time updates on public transport schedules, weather conditions, and even emergency alerts. Smart healthcare solutions are being integrated to monitor public health metrics, offering data that can help prevent disease outbreaks and improve medical response times. Waste management, as outlined earlier, is enhanced through IoT by ensuring timely collection and reducing litter. This connectivity extends to public safety as well, where smart surveillance systems and IoT-based emergency response services significantly enhance urban safety. The cumulative effect of these innovations contributes to creating more livable, resilient, and forward-thinking urban environments.

SMART STREETLIGHTS

Smart streetlights are a pivotal component of the smart city infrastructure, designed to enhance urban lighting efficiency while reducing energy consumption and maintenance costs. These innovative lighting systems employ sensors, wireless connectivity, and adaptive controls to adjust lighting based on real-time conditions such as ambient light levels, pedestrian activity, and traffic flow. By implementing smart streetlights, cities can improve public safety, lower their carbon footprint, and optimize maintenance processes. Here are five examples of smart streetlight applications, along with their costs and benefits:

1. **Adaptive Lighting Systems:**

Cost: Usually range from $1,000 to $5,000 per unit, based on sensor capabilities and control software.

Benefit: Adjusts lighting intensity based on surrounding conditions, reducing unnecessary energy use and enhancing visibility when needed, leading to significant energy savings.

2. **Remote Monitoring and Control:**

Cost: Typically between $500 and $2,000 per light, depending on connectivity and management features.

Benefit: Allows for real-time monitoring and control of streetlight operations, enabling quick responses to faults or outages and reducing maintenance time and costs.

3. **Presence Detection Sensors:**

Cost: Generally ranges from $200 to $1,000 per sensor, influenced by detection range and technology used.

Benefit: Activates lighting when pedestrian or vehicle movement is detected, improving safety and reducing energy use during low-activity periods.

4. **Solar-Powered Streetlights:**

Cost: Usually between $2,000 and $7,000 per unit, depending on solar panel efficiency and battery storage capacity.

Benefit: Utilizes renewable energy to power streetlights, decreasing reliance on the electrical grid and lowering overall operational costs while promoting sustainability.

5. **Integrated Air Quality Sensors:**

Cost: Typically ranges from $500 to $3,000 per sensor, based on sensitivity and data transmission capabilities.

Benefit: Monitors local air quality in real-time, providing valuable data for environmental management and improving public health by identifying pollution hotspots.

TRAFFIC MANAGEMENT SYSTEMS

As smart city technologies become increasingly integrated into urban infrastructure, the transformation of cities will be profound and multifaceted. The deployment of IoT applications, such as smart streetlights and advanced traffic management systems, will lead to more efficient energy use, reduced traffic congestion, and enhanced public safety. These innovations will not only provide real-time solutions to common urban issues but also contribute to long-term sustainability by lowering carbon emissions and promoting renewable energy usage. The increased connectivity facilitated by IoT will empower citizens with accessible information and services, promoting higher levels of engagement and inclusivity. Additionally, improved data collection and analysis will enable better-informed decision-making by city planners and officials, leading to a more responsive and adaptive urban environment. Overall, these changes will create smarter, more livable cities that can dynamically respond to the needs of their inhabitants while fostering economic growth and sustainability.

WASTE MANAGEMENT SOLUTIONS

The integration of advanced IoT applications within smart city frameworks will revolutionize urban living, creating more efficient, sustainable, and responsive environments. By implementing smart streetlights, adaptive traffic management systems, and intelligent waste management solutions, cities can significantly reduce energy consumption, cut operational costs, and enhance public safety. These technologies enable real-time monitoring and dynamic adjustments, leading to optimized resource usage and reduced environmental impact. Furthermore, improved data collection allows for informed decision-making, helping city planners address issues promptly and effectively. This holistic approach fosters a higher quality of life for residents, promoting economic development while ensuring long-term sustainability and resilience in the face of growing urban challenges.

PUBLIC SAFETY AND SECURITY ADVANCEMENTS

IoT will significantly help cities improve public safety and security by employing advanced technologies such as smart surveillance systems, real-time crime monitoring, and predictive analytics. For instance, IoT-enabled cameras and sensors can provide continuous monitoring of public spaces, identifying unusual activity and alerting authorities to potential threats in real-time. This rapid response capability reduces crime rates and enhances the overall perception of safety among residents. Additionally, emergency response services can leverage IoT data to optimize resource allocation and arrival times, ensuring that help is dispatched more efficiently and effectively. By integrating these intelligent solutions, cities can build a more secure and resilient urban environment that proactively addresses safety concerns and protects its inhabitants.

SUMMARY

This chapter delved into the transformative impact of IoT applications on urban environments, highlighting several key areas of innovation and their benefits. Smart streetlights, with their adaptive lighting systems, remote monitoring capabilities, and energy-efficient designs, were shown to enhance public safety, reduce energy consumption, and lower maintenance costs. Advanced traffic management systems were discussed as tools for easing congestion and improving urban mobility, further contributing to sustainability goals. The chapter also explored the integration of intelligent waste management solutions, which optimize resource usage and reduce environmental impact. Finally, advancements in public safety and security through IoT-enabled surveillance and real-time crime monitoring were examined, illustrating how these technologies can enhance the overall quality of life. By implementing these smart technologies, cities can become more efficient, sustainable, and responsive to the needs of their inhabitants.

CONCLUSION

Future Trends in IoT

As we look towards the future, the Internet of Things (IoT) is poised to drive even more groundbreaking innovations across various sectors. One significant trend is the continued development of AI-powered IoT devices. By integrating artificial intelligence, IoT systems will become more autonomous and capable of making more accurate predictions, decisions, and actions without human intervention. This advancement will be particularly transformative in healthcare, where AI-powered IoT devices can monitor patients in real-time, predict health issues before they become severe, and even suggest personalized treatment plans.

Another notable trend is the expansion of IoT in smart homes, where everyday devices such as refrigerators, thermostats, and even ovens are becoming increasingly connected and capable of communicating with each other. This interconnectivity not only offers convenience but also contributes to energy efficiency as these devices can optimize their operation based on real-time data. In industrial settings, 5G will facilitate more precise and efficient operations, predictive maintenance, and enhanced worker safety through connected sensors and devices.

Blockchain technology is also expected to play a crucial role in securing IoT ecosystems. By providing a decentralized and tamper-proof ledger, blockchain can address some of the significant security concerns associated with IoT, such as data breaches and unauthorized access. This will be particularly important for critical infrastructures like smart grids and healthcare systems, where data integrity and security are paramount.

Conclusion

IoT will continue to drive global transformation, creating a more intelligent, interconnected, and efficient world. The integration of artificial intelligence, 5G technology, blockchain, and edge computing will elevate IoT to new heights, unlocking potential across various sectors and improving quality of life. As IoT evolves, its impact will be felt broadly—from revolutionizing healthcare and agriculture to enhancing industrial operations and urban living. The relentless advancement of IoT technologies ensures a future where data-driven insights lead to more informed decisions, streamlined processes, and innovative solutions to pressing challenges. Ultimately, IoT stands as a cornerstone of technological progress, heralding an era of unprecedented growth and connectivity that will shape the world for generations to come.

REFERENCES

Gill, K. S., Yang, S. H., Yao, F., & Lu, X. (2012). A ZigBee-based home automation system. *IEEE Transactions on Consumer Electronics*, 55(2), 422-430.

Al-Momani, A. M., Lakany, H., & McMeekin, D. A. (2018). Smart home automation system for elderly, and handicapped people using Xbee. *Proceedings of the 13th International Conference on Mobile Systems and Pervasive Computing (MobiSPC)*, 84-91.

Altun, K. P., & Kocakulak, M. (2017). Real-time smart house application via a secure communication using GSM. *Computers & Electrical Engineering*, 57, 99-111.

Kumar, R., Maity, M., & Pal, S. (2016). Smart home automation based on Internet of Things. *Proceedings of the 2nd International Conference on Computer and Information Technology (ICCIT)*, 1-4.

Wang, S., Yang, X., Lü, Z., & Xu, J. (2012). Smart Home System Based on IOT Technologies. *Procedia Engineering*, 15, 2087-2092.

Lee, J., Bagheri, B., & Kao, H. A. (2015). A cyber-physical systems architecture for industry 4.0-based manufacturing systems. *Manufacturing Letters*, 3, 18-23.

Gilchrist, A. (2016). *Industry 4.0: The Industrial Internet of Things*. Apress.

Delsing, J. (2017). *IoT automation: Arrowhead framework*. CRC Press.

Dutton, W. H. (2014). Putting things to work: Social and policy challenges for the Internet of Things. *info*, 16(3), 1-21.

Wan, J., Yan, H., Suo, H., & Li, F. (2011). Advances in Cyber-Physical Systems Research. *Proceedings of the 10th International Conference on Computer Applications and Industrial Electronics (ICCAIE)*, 589-595.

Islam, S. M. R., Kwak, D., Kabir, M. H., Hossain, M., & Kwak, K. S. (2015). The Internet of Things for Health Care: A Comprehensive Survey. *IEEE Access*, 3, 678-708.

Verma, P., Sood, S. K., & Kalra, S. (2018). A Comprehensive Review of Devices and Techniques Available for Implementation of IoT-Based Healthcare Systems. *Journal of Medical Systems*, 42(2), 1-15.

Amancio, D. R., Oliveira, O. N., & Costa, L. F. (2014). A systematic analysis of the literature and microarray data on the impact of IoT in healthcare. *PLoS ONE*, 9(9), e106888.

Darshan, K. R., & Anandakumar, K. R. (2015). A comprehensive review on usage of Internet of Things (IoT) in healthcare system. *2015 International Conference on Emerging Research in Electronics, Computer Science and Technology (ICERECT)*, 132-136.

Kovatsch, M., Mayer, S., & Shelby, Z. (2015). Practical semantics for the IoT: Contributing to the Web of Things. *IEEE Internet Computing*, 19(2), 10-18.

Zhang, L., Shen, Q., & Chen, Z. (2016). Internet of Things Research Framework for Intelligent Agriculture. *Journal of Software Engineering and Applications*, 5(07), 384-396.

Li, Z., Dappa, J., & Lehmann, T. (2017). An IoT-enabled smart watering system for increasing efficiency of agriculture. *40th International Conference on Telecommunications and Signal Processing (TSP)*, 730-733.

Chakraborty, T., Pande, U. S., & Bajaj, S. (2019). Internet of Things (IoT) in Agriculture. *IEEE Conference on Emerging Devices and Smart Systems (ICEDSS)*, 253-255.

Demestichas, K., Peppas, K., & Alexakis, E. (2018). Agricultural and food systems in 2050: a report of the Foresight Committee Ymini-IOT platform in agriculture sector. *Computers and Electronics in Agriculture*, 155, 13-18.

Kamilaris, A., Kartakoullis, A., & Prenafeta-Boldú, F. X. (2017). A review on the practice of big data analysis in agriculture. *Computers and Electronics in Agriculture*, 143, 23-37.

Grewal, D., Roggeveen, A. L., & Nordfält, J. (2017). The Future of Retailing. *Journal of Retailing*, 93(1), 1-6.

Atzori, L., Iera, A., & Morabito, G. (2010). The Internet of Things: A survey. *Computer Networks*, 54(15), 2787-2805.

Pantano, E., & Priporas, C. V. (2016). The effect of mobile retailing technology on consumer choice behavior: Insights from an empirical study. *Journal of Retailing and Consumer Services*, 30, 198-204.

Kim, M. J. (2018). Customer tracking and shopping behavior analytics using IoT technologies in retailing. *Business Process Management Journal*, 24(4), 970-983.

Brynjolfsson, E., Hu, Y. J., & Rahman, M. S. (2013). Competing in the Age of Omnichannel Retailing. *MIT Sloan Management Review*, 54(4), 23-29.

Madhu-N, S., & Balaraman, K. (2016). IoT for energy efficiency in buildings. *2016 International Symposium on Green Buildings and Sustainable ICT*, 243-250.

Drgoňa, J., Picard, D., Kvasnica, M., & Helsen, L. (2020). Distributed predictive control for smart energy systems: Overview and application. *Renewable and Sustainable Energy Reviews*, 108, 155-196.

Khan, R., Khan, S. U., Zaheer, R., & Khan, S. (2012). Future Internet: The Internet of Things Architecture, Possible Applications and Key Challenges. *Proceedings of the 10th International Conference on Frontiers of Information Technology*, 257-260.

Chiang, M., Niyato, D., & Wang, P. (2016). Internet of Things (IoT): Architecture, protocols and services. *Wireless Networks*, 22(1), 41-59.

Aung, Z., Yeoh, W., Chan, S. H., & Fong, K. Y. (2012). Energy efficiency in green data centres: A research review. *Renewable and Sustainable Energy Reviews*, 16(6), 4105-4118.

Anjomshoaa, A., Duarte, F., Rennings, D., Faro, A., & Ratti, C. (2018). Transker: A crowd-sourced system for IoT-based transport infrastructure monitoring. *Journal of Big Data*, 5, 31.

Barbaresso, J. C., Cordahi, G., & Garcia, D. (2015). Automated Vehicle Research for Enhanced Safety. *Transportation Research Record*, 2530, 17-24.

Alam, S., & Ferreira, L. (2013). Intelligent Transport Systems: Emerging technologies and applications. *IET Intelligent Transport Systems*, 7(3), 247-255.

Lin, J., Rivano, H., & Le Mouel, F. (2017). A survey of smart transportation solutions based on Internet of Things. *IEEE Internet of Things Journal*, 4(1), 4-12.

Gerla, M., Lee, E.-K., Pau, G., & Lee, U. (2014). Internet of Vehicles: From intelligent grid to autonomous cars and vehicular clouds. *IEEE World Forum on Internet of Things (WF-IoT)*, 241-246.

Giffinger, R., Fertner, C., Kramar, H., & Kalasek, R. (2007). Smart cities: Ranking of European medium-sized cities. *Centre of Regional Science (SRF), Vienna University of Technology*.

Batty, M., Axhausen, K. W., Giannotti, F., Pozdnoukhov, A., Bazzani, A., Wachowicz, M., Ouzounis, G., & Portugali, Y. (2012). Smart cities of the future. *European Physical Journal Special Topics*, 214, 481-518.

Zygiaris, S. (2013). Smart City Reference Model: Assisting planners to conceptualize the building of smart city innovation ecosystems. *Journal of the Knowledge Economy*, 4, 217-231.

Schaffers, H., Komninos, N., Pallot, M., Trousse, B., Nilsson, M., & Oliveira, A. (2011). Smart Cities and the Future Internet: Towards Cooperation Frameworks for Open Innovation. *The Future Internet Assembly*, 431-446.

Anthopoulos, L. G., & Fitsilis, P. (2010). From digital to ubiquitous cities: Defining a common architecture for urban development. *2010 Sixth International Conference on Intelligent Environments*, 301-306.

Tang, J., & McLoughlin, C. (2021). Implementing Internet of Things in Education: Perspectives, Challenges, and Opportunities in Smart Learning Environments. *SpringerBriefs in Education*.

Alamri, A., Hassan, M. M., Alnuem, M. A., & Hossain, M. S. (2014). A survey on sensor-cloud: Architecture, applications, and approaches. *International Journal of Distributed Sensor Networks*, 10(2), 917923.

Pham, X., & Pham, T. H. (2021). Using IoT and Big Data for e-learning systems. *International Journal of Emerging Technologies in Learning (iJET)*, 16(24), 98-113.

Güngör, V. C., & Hancke, G. P. (2009). Industrial wireless sensor networks: Challenges, design principles, and technical approaches. *IEEE Transactions on Industrial Electronics*, 56(10), 4258-4265.

Mehdiyev, N., & Osmanov, V. (2020). Implementing IoT and AI in Digital Learning Environment. *2020 IEEE International Conference on Smart Information Systems and Technologies (SIST)*, 29-34.

Ma, Y., Richards, M., Ghanem, M., Ketley, C., & Guo, Y. (2008). Mapping Climate Data in the GRID. *Phil. Trans. R. Soc. A*, 366, 3305-3317.

Chhetri, M. B., & Chetty, M. (2019). Environment Monitoring Using IoT and Multi-sensor Data Fusion Techniques. *Future Generation Computer Systems*, 93, 290-299.

Hart, J. K., & Martinez, K. (2006). Environmental Sensor Networks: A revolution in the earth system science? *Earth-Science Reviews*, 78(3-4), 177-191.

Madeira, J., & Leitao, P. (2017). Review on IoT-based data aquisition methods for environmental monitoring. *IFAC-PapersOnLine*, 50(1), 4186-4191.

Wawerla, S., & Lempert, C. (2019). Adaptive Environmental Monitoring Deployments Based on IoT Technologies. *International Journal of Communication Systems*, 32(14), e4116.

Weber, R. H. (2010). Internet of Things – New security and privacy challenges. *Computer Law & Security Review*, 26(1), 23-30.

Roman, R., Zhou, J., & Lopez, J. (2013). On the features and challenges of security and privacy in distributed internet of things. *Computer Networks*, 57(10), 2266-2279.

Sicari, S., Rizzardi, A., Grieco, L. A., & Coen-Porisini, A. (2015). Security, privacy and trust in Internet of Things: The road ahead. *Computer Networks*, 76, 146-164.

Xu, L. D., He, W., & Li, S. (2014). Internet of Things in industries: A survey. *IEEE Transactions on Industrial Informatics*, 10(4), 2233-2243.

Zheng, Y., & Vasconcelos, W. (2020). The impact of IoT on security and privacy: A review of post-GDPR challenges. *Future Internet*, 12(12), 221.